Landforms of the Colorado Plateau

The Colorado River has sliced deeply into an astonishing array of rock strata within the Colorado Plateau. The result is an exhibition of color and form unrivaled by any other province of the planet.

We tiptoe through fragments of time —
yesterday a dune,
today a soaring arch,
tomorrow a sandy hillock
of tumbled boulders.

"Landforms—Heart of the Colorado Plateau" *encompasses the canyons and mesas along the Colorado River from Arches National Park downriver to Lees Ferry, and from Bryce Canyon National Park east to Natural Bridges National Monument.*

Front cover: Knobs of eroded sandstone reveal a record of marching sand dunes and fickle winds. Inside front cover: Balanced Rock, Arches National Park, Utah. Page 1: Erosion etches Navajo Sandstone rimrock of the Vermilion Cliffs, Arizona. Pages 2/3: Guy's Eye near Lake Powell, Utah. Pages 4/5: Rock mirrors the form of flowing water in a slot canyon near Page, Arizona.

Edited by Mary L. Van Camp.
Book design by K. C. DenDooven.

Second Printing, 1998

LANDFORMS—HEART OF THE COLORADO PLATEAU:
The Story Behind the Scenery

LC 95-75094. ISBN 0-88714-090-4.

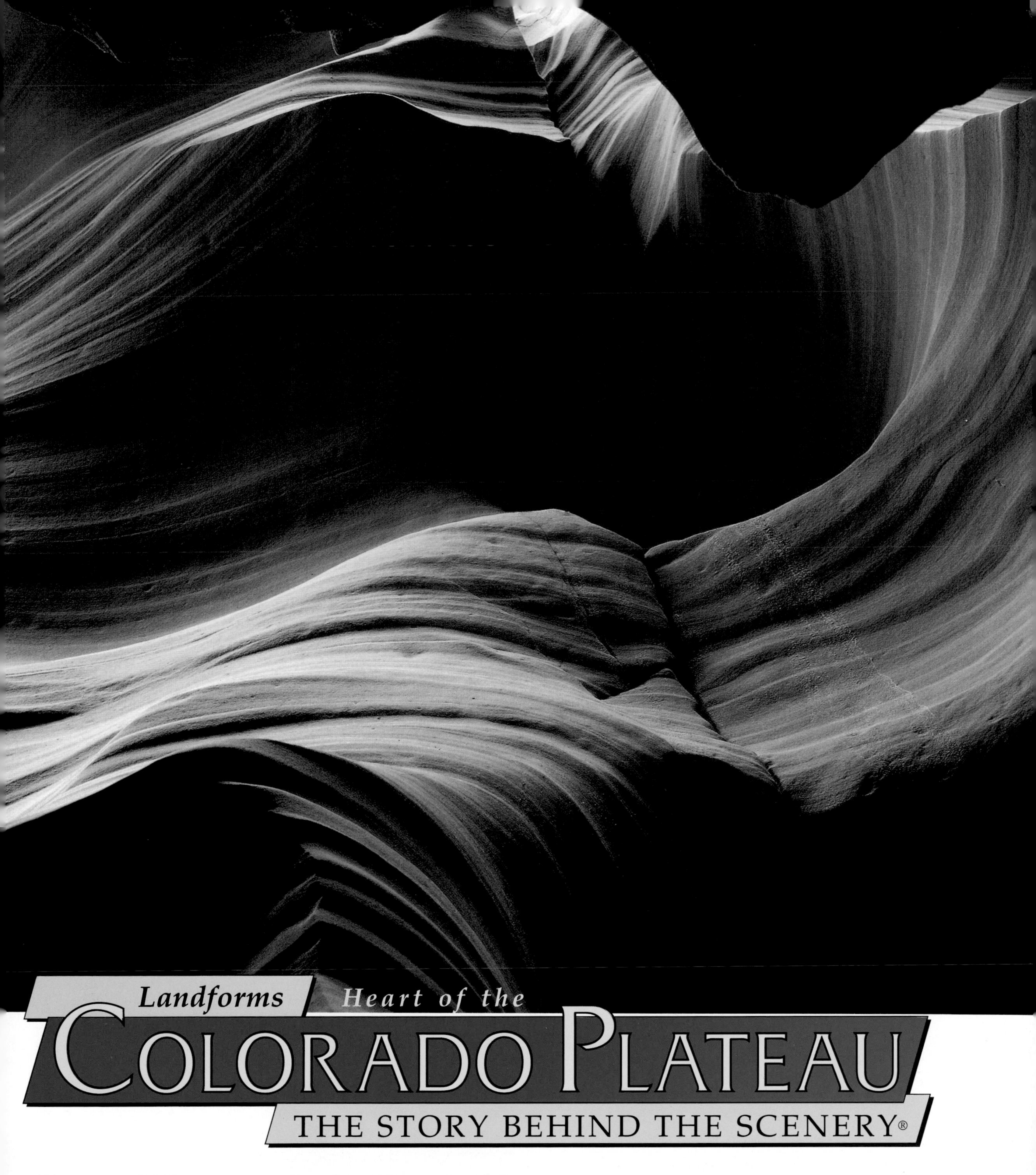

Landforms

Heart of the COLORADO PLATEAU

THE STORY BEHIND THE SCENERY®

Text and Photography by Gary Ladd

Gary Ladd has explored, photographed, and studied the geology of the Colorado Plateau for over 25 years. He is a Museum of Northern Arizona Ventures trip leader, and Sierra Club wilderness backpack leader.

Technical assistance provided by Frank Howd Ph.D., retired geology professor, University of Maine.

Look! There it lies — bare rock cleaved by precipitous canyons, thousands and thousands of square miles of rock and stone exposed to our view. The sight is stirring, and the landscape is a delight for both the eye and the intellect.

The naked rock whispers and hints at the geologic story of the Lake Powell region. Murmurs of ancient oceans, deserts, and rivers issue from the cliffs, canyons, and alcoves.

What happened here? How did this maze of canyons come to be? How old are the canyons? What other landforms and environments occupied this region?

Geology, the study of the Earth, opens windows from which we can glimpse the past through the landforms of the present. The view answers many of our questions about the bare rocks and deep canyons.

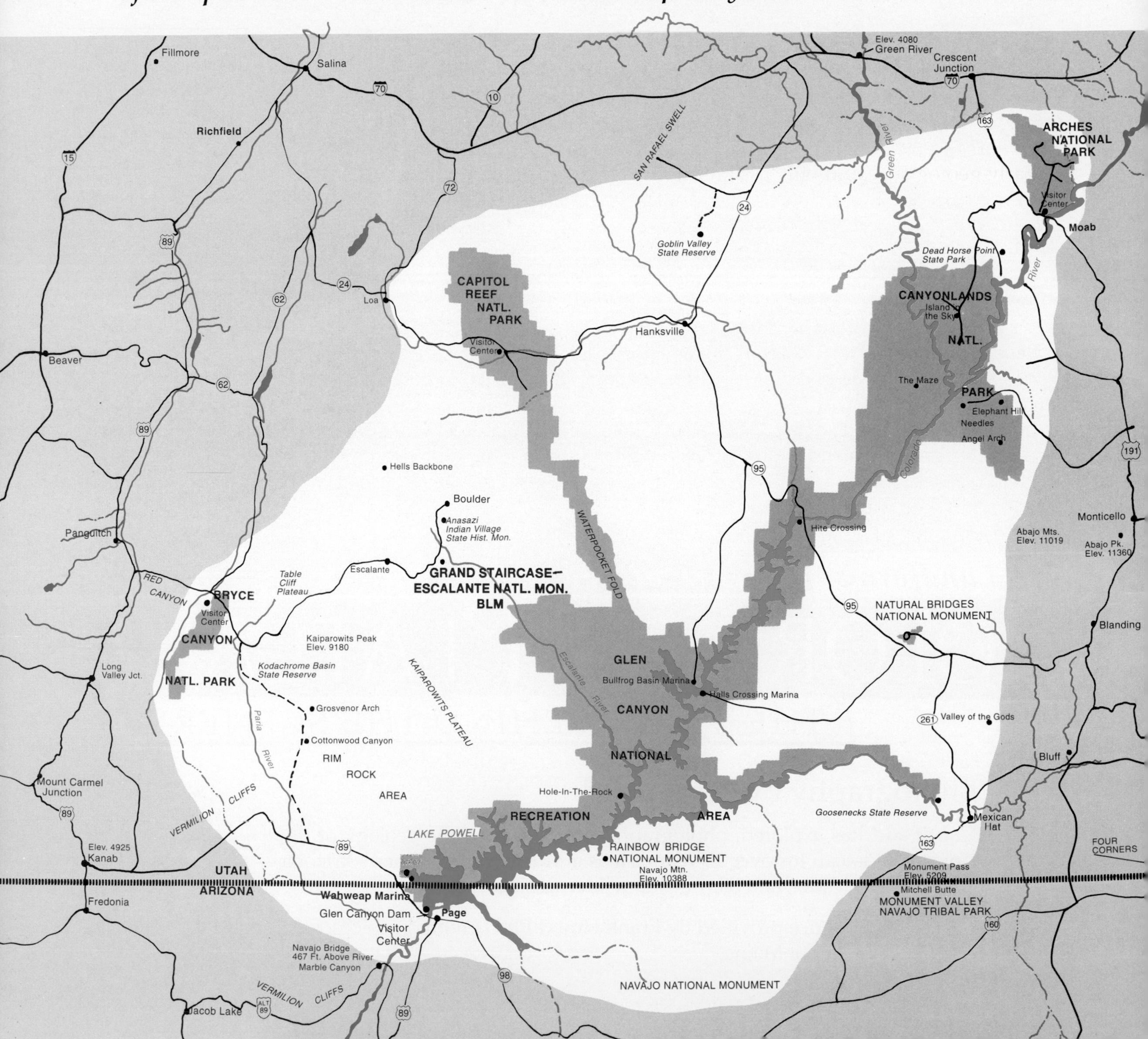

Landforms: The story of how landscapes, seashores, rivers, lakes, plains, canyons, wetlands, deserts, and mountains were created and the sequence of geologic events, chronicled in the rocks, that culminated in these terrains on which we walk today.

Yesterday and the Time Line

In the beginning the Sun, Earth, and other planets were born from "star dust," the detritus flung from the hulks of exhausted early stars. At first the Earth was a hellish place, meteor and planetesimal impacts were frequent, its rocks often molten, its oceans ephemeral. Eventually, over the course of eons, our planet matured into an oasis in the immensity of space.

The moderating conditions allowed life to gain a toehold and ultimately blossom and diversify. Yet, enormous amounts of time would elapse before any being would notice a pretty rock, pick it up and ask, "Where did this come from? Is it a thousand years old? Ten thousand years? Even older?"

Early geologists could only assign *relative* ages to rock units. A particular layer might be considered older than another because it lay deeper in the Earth. Thus rock strata were placed in classes that by stratigraphic position, composition or fossil remains hinted that they were closely related. These relative age groupings were often named after the area in which they were well exposed—the Permian Period for Perm in Russia, the Cambrian Period for Cambria or Wales.

Several methods were devised to reveal the *absolute* ages of rock units. But the answers never agreed and the methods were later found to be flawed. Then came a breakthrough.

In 1896, radioactivity was discovered. The spontaneous decay of some isotopes possessed the steady, long-running "clock" upon which the geologic calendar could be based. The results were humbling: Geologists learned that their carefully constructed relative age calendar spanned only the final 12 percent of geologic time. By using a combination of relative ages and the new radiometric dating method, most rocks were determined to be not thousands or even millions of years old, they were often tens to hundreds of millions of years old or older still. And Earth, as has more recently been discovered, began its career about 4.6 billion years ago.

The Heart of the Colorado Plateau Time Line picks up the trail of geologic history 285 million years ago. Ninety-four percent of Earth's history has already sneaked by. The stage is set for a pageant of events and affairs chronicled in the strata that will sweep onward to the development of the Colorado River, the cleaving of its deepest canyons, and the creation of Delicate Arch.

Hermosa group — *Halgaito (Elephant Canyon) formation* —?—

PENNSYLVANIAN PERIOD — PERMIAN PERIOD →

285 MYA (Millions of Years Ago) — 280 MYA

← PALEOZOIC ERA →

A slab of Wingate Sandstone recently tumbled from an overhanging cliff to shatter on a bench below. What made the slab fall? Frost wedging, most likely, the familiar but powerful expansive force of water as it turns to ice on a freezing winter night.

Time and Change

Imagine this — the Earth without human inhabitants. There would be no roads, no towns or cities, no power lines, no ships upon the sea. Oceans would pound and rivers flow, but millennia would pass unnoticed. Without clocks or calendars time may seem meaningless or, perhaps, irrelevant. Yet, the vast bulk of Earth's history lapsed before the first human eyes blinked open.

Even more astonishing, Earth's continents were barren of life until rather recently. Eighty percent of Earth's history passed before the first plant took root on dry land; 90 percent passed before animals walked the hills and plains. We upstarts have a sadly unrealistic concept of time and history!

Cedar Mesa formation of Cutler group ——— ?—— *Organ Rock formation of*

PERMIAN PERIOD →

275 MYA

Much of the action of canyon cutting occurs during inclement weather. Heavy rains, high winds, and cold snaps all accelerate the pace of geologic change. An afternoon of high winds will chisel away at rock outcrops grain by grain. The waste rock is tossed into ravines and washes. A brief summer thunderstorm will snatch up the sand and carry it off toward the Colorado River. Along the way the grit-laden waters scour the bedrock. Where the bedrock has been weakened by fractures, the rasping water cuts even more effectively.

THE POWERS OF WATER AND TIME

Consider the powers of water and time. The rock of the Earth is for the greatest part highly durable. The so-called everlasting mountains are composed of rock. Yet, solid rock is no match for the tireless work of water. Water is abundant and it moves about the planet with impunity. Water dissolves and erodes. Water makes mush of solid rock, given enough time.

Time, the second power, is more abundant but less obvious. Only time allows water to assume its role as an important geologic power. Time has been available in vast and noble quantities, time measured in tens and hundreds of millions of years. Deep time.

Simply said, things change. Given enough time things change a great deal — mountains thrust skyward, buttes are cut from plateaus, seas come and go, canyons slice into the land. On a geologic time scale mountains and mesas are as insubstantial as clouds; rivers are as vaporous as mist.

Cutler group ——— ?— *White Rim formation of Cutler group* ——— ?—

270 MYA **265 MYA**

Bridges are the artwork of streams. Arches often develop where only the more subtle forces prevail—weathering, frost wedging and spalling—without the aid of a vigorous stream. Sipapu Bridge in Natural Bridges National Monument, Utah, leaps 268 feet across White Creek.

Yet, geologic change is rarely catastrophic. Only volcanic eruptions and a few other extraordinary events violate the rule. (For instance, there is growing evidence of rare impacts by large asteroids or comets. The impacts have devastating local geologic effects but, more importantly, disasterous global environmental results that drastically alter the course of biologic evolution.) Most of geologic change is tediously slow. Not surprisingly, early geologists found this concept difficult to swallow. Our view of the pace of geologic evolution is myopic. As living beings with lives of pitifully short duration, we simply don't exist long enough to detect long-term trends. Lucretius wrote two thousand years ago: "The first beginnings of things cannot be distinguished by the eye."

The transition from one past environment (say, an ocean) to another (say, a desert) was imperceptible. The pace of change then was as it is now . . . exceedingly slow. All that is necessary is time, just time.

Geologic Time

Let's assemble a list of benchmark events appropriate to the scale of geologic time: The age of the universe is believed to be about 15 billion years, the age of the Earth about 4.6 billion years, the oldest Grand Canyon rocks approach 2 billion years, the age of Lake Powell region rocks range from about 100 to 300 million years, and the age of the Colorado River canyons is about 5.5 million years.

Look carefully at these numbers. Note that the canyons — those features one may be likely to equate with great age — are far *younger* than the rocks through which they slice. Furthermore, the rocks are but a fraction of the age of the Earth. This is an important point: In spite of their ancient appearance, the canyons of the Colorado River are merely a shade over one-tenth of one percent of the age of the Earth.

The Colorado River (1,100 feet below the rim of Glen Canyon) sweeps through a meander toward Lees Ferry, Arizona. A few miles downstream from the bend, the river breaks through the Echo Cliffs Monocline. There, Glen Canyon abruptly ends when older, harder rocks rise to river level.

Kaibab Limestone formation ≡?≡ *x* *x* *x*

PERMIAN PERIOD →

260 MYA

x=Significant Periods of Erosion or Non-Deposition

x
x
x
x
x
255 MYA
250 MYA

x x x x ≡?≡ *Moenkopi formation*

PERMIAN PERIOD → | TRIASSIC PERIOD →

245 MYA

MESOZOIC ERA →

Landforms: The story of how landscapes, seashores, rivers, lakes, plains, canyons, wetlands, deserts, and mountains were created and the sequence of geologic events, chronicled in the rocks, that culminated in these terrains on which we walk today.

Yesterday and the Time Line

In the beginning the Sun, Earth, and other planets were born from "star dust," the detritus flung from the hulks of exhausted early stars. At first the Earth was a hellish place, meteor and planetesimal impacts were frequent, its rocks often molten, its oceans ephemeral. Eventually, over the course of eons, our planet matured into an oasis in the immensity of space.

The moderating conditions allowed life to gain a toehold and ultimately blossom and diversify. Yet, enormous amounts of time would elapse before any being would notice a pretty rock, pick it up and ask, "Where did this come from? Is it a thousand years old? Ten thousand years? Even older?"

Early geologists could only assign *relative* ages to rock units. A particular layer might be considered older than another because it lay deeper in the Earth. Thus rock strata were placed in classes that by stratigraphic position, composition or fossil remains hinted that they were closely related. These relative age groupings were often named after the area in which they were well exposed—the Permian Period for Perm in Russia, the Cambrian Period for Cambria or Wales.

Several methods were devised to reveal the *absolute* ages of rock units. But the answers never agreed and the methods were later found to be flawed. Then came a breakthrough.

In 1896, radioactivity was discovered. The spontaneous decay of some isotopes possessed the steady, long-running "clock" upon which the geologic calendar could be based. The results were humbling: Geologists learned that their carefully constructed relative age calendar spanned only the final 12 percent of geologic time. By using a combination of relative ages and the new radiometric dating method, most rocks were determined to be not thousands or even millions of years old, they were often tens to hundreds of millions of years old or older still. And Earth, as has more recently been discovered, began its career about 4.6 billion years ago.

The Heart of the Colorado Plateau Time Line picks up the trail of geologic history 285 million years ago. Ninety-four percent of Earth's history has already sneaked by. The stage is set for a pageant of events and affairs chronicled in the strata that will sweep onward to the development of the Colorado River, the cleaving of its deepest canyons, and the creation of Delicate Arch.

Hermosa group — *Halgaito (Elephant Canyon) formation* — ?

PENNSYLVANIAN PERIOD — PERMIAN PERIOD →

285 MYA (Millions of Years Ago) — **280 MYA**

← PALEOZOIC ERA →

A slab of Wingate Sandstone recently tumbled from an overhanging cliff to shatter on a bench below. What made the slab fall? Frost wedging, most likely, the familiar but powerful expansive force of water as it turns to ice on a freezing winter night.

Time and Change

Imagine this — the Earth without human inhabitants. There would be no roads, no towns or cities, no power lines, no ships upon the sea. Oceans would pound and rivers flow, but millennia would pass unnoticed. Without clocks or calendars time may seem meaningless or, perhaps, irrelevant. Yet, the vast bulk of Earth's history lapsed before the first human eyes blinked open.

Even more astonishing, Earth's continents were barren of life until rather recently. Eighty percent of Earth's history passed before the first plant took root on dry land; 90 percent passed before animals walked the hills and plains. We upstarts have a sadly unrealistic concept of time and history!

Cedar Mesa formation of Cutler group ——— ?——— *Organ Rock formation of*

PERMIAN PERIOD ⟶

275 MYA

x x x x ?Moenkopi formation

PERMIAN PERIOD → | TRIASSIC PERIOD →

245 MYA

MESOZOIC ERA →

255 MYA
250 MYA

Look! There it lies — bare rock cleaved by precipitous canyons, thousands and thousands of square miles of rock and stone exposed to our view. The sight is stirring, and the landscape is a delight for both the eye and the intellect.

The naked rock whispers and hints at the geologic story of the Lake Powell region. Murmurs of ancient oceans, deserts, and rivers issue from the cliffs, canyons, and alcoves.

What happened here? How did this maze of canyons come to be? How old are the canyons? What other landforms and environments occupied this region?

Geology, the study of the Earth, opens windows from which we can glimpse the past through the landforms of the present. The view answers many of our questions about the bare rocks and deep canyons.

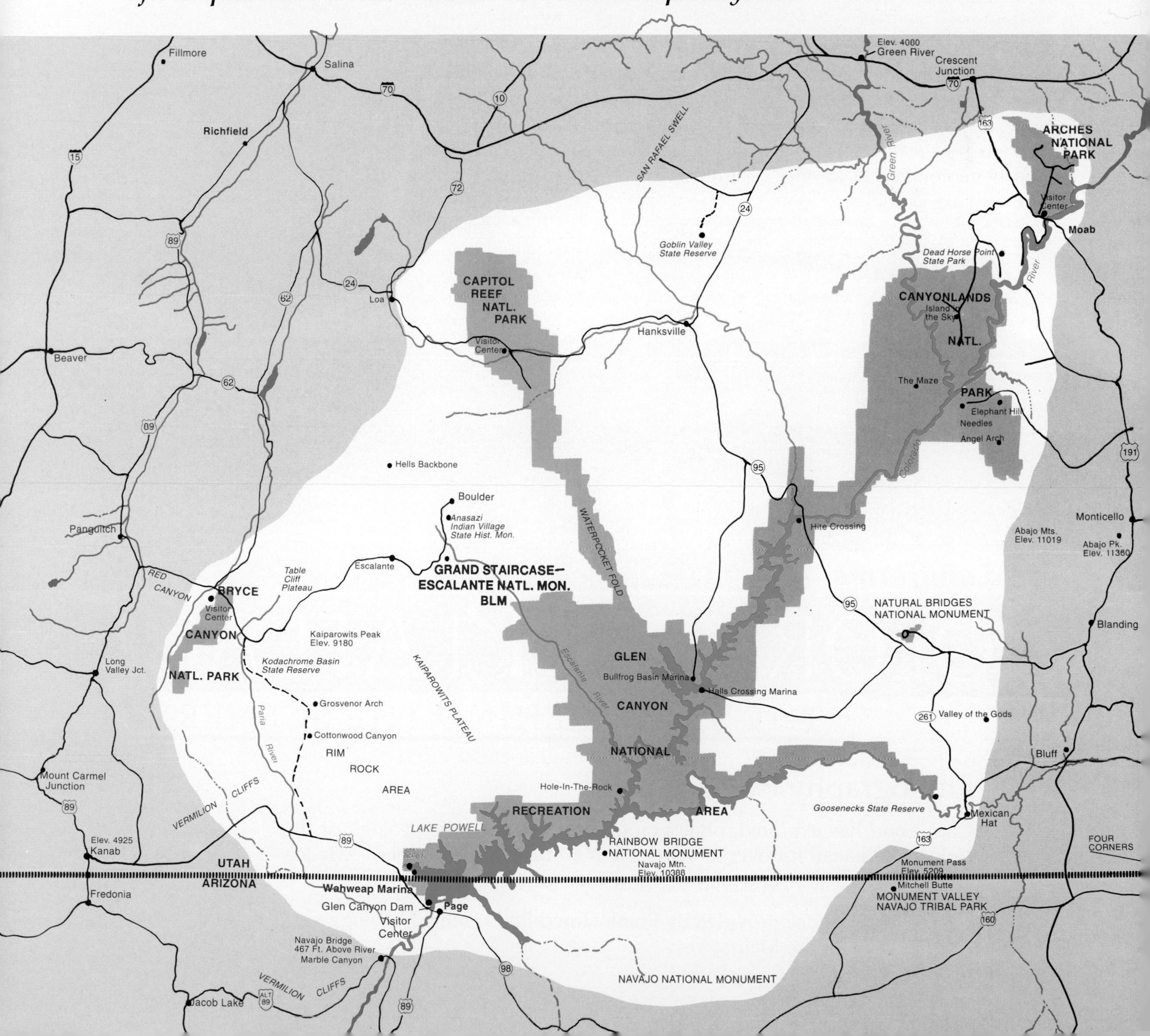

Landforms Heart of the

COLORADO PLATEAU

THE STORY BEHIND THE SCENERY®

Text and Photography by Gary Ladd

Gary Ladd has explored, photographed, and studied the geology of the Colorado Plateau for over 25 years. He is a Museum of Northern Arizona Ventures trip leader, and Sierra Club wilderness backpack leader.

Technical assistance provided by Frank Howd Ph.D., retired geology professor, University of Maine.

Bridges are the ▶ artwork of streams. Arches often develop where only the more subtle forces prevail—weathering, frost wedging and spalling—without the aid of a vigorous stream. Sipapu Bridge in Natural Bridges National Monument, Utah, leaps 268 feet across White Creek.

Yet, geologic change is rarely catastrophic. Only volcanic eruptions and a few other extraordinary events violate the rule. (For instance, there is growing evidence of rare impacts by large asteroids or comets. The impacts have devastating local geologic effects but, more importantly, disasterous global environmental results that drastically alter the course of biologic evolution.) Most of geologic change is tediously slow. Not surprisingly, early geologists found this concept difficult to swallow. Our view of the pace of geologic evolution is myopic. As living beings with lives of pitifully short duration, we simply don't exist long enough to detect long-term trends. Lucretius wrote two thousand years ago: "The first beginnings of things cannot be distinguished by the eye."

The transition from one past environment (say, an ocean) to another (say, a desert) was imperceptible. The pace of change then was as it is now . . . exceedingly slow. All that is necessary is time, just time.

Geologic Time

Let's assemble a list of benchmark events appropriate to the scale of geologic time: The age of the universe is believed to be about 15 billion years, the age of the Earth about 4.6 billion years, the oldest Grand Canyon rocks approach 2 billion years, the age of Lake Powell region rocks range from about 100 to 300 million years, and the age of the Colorado River canyons is about 5.5 million years.

Look carefully at these numbers. Note that the canyons — those features one may be likely to equate with great age — are far *younger* than the rocks through which they slice. Furthermore, the rocks are but a fraction of the age of the Earth. This is an important point: In spite of their ancient appearance, the canyons of the Colorado River are merely a shade over one-tenth of one percent of the age of the Earth.

The Colorado River (1,100 feet below the rim of Glen ▶ Canyon) sweeps through a meander toward Lees Ferry, Arizona. A few miles downstream from the bend, the river breaks through the Echo Cliffs Monocline. There, Glen Canyon abruptly ends when older, harder rocks rise to river level.

Kaibab Limestone formation ——— ≡?≡ x x x

——— PERMIAN PERIOD ———▶

260 MYA

x=Significant Periods of Erosion or Non-Deposition

Much of the action of canyon cutting occurs during inclement weather. Heavy rains, high winds, and cold snaps all accelerate the pace of geologic change. An afternoon of high winds will chisel away at rock outcrops grain by grain. The waste rock is tossed into ravines and washes. A brief summer thunderstorm will snatch up the sand and carry it off toward the Colorado River. Along the way the grit-laden waters scour the bedrock. Where the bedrock has been weakened by fractures, the rasping water cuts even more effectively.

The Powers of Water and Time

Consider the powers of water and time. The rock of the Earth is for the greatest part highly durable. The so-called everlasting mountains are composed of rock. Yet, solid rock is no match for the tireless work of water. Water is abundant and it moves about the planet with impunity. Water dissolves and erodes. Water makes mush of solid rock, given enough time.

Time, the second power, is more abundant but less obvious. Only time allows water to assume its role as an important geologic power. Time has been available in vast and noble quantities, time measured in tens and hundreds of millions of years. Deep time.

Simply said, things change. Given enough time things change a great deal — mountains thrust skyward, buttes are cut from plateaus, seas come and go, canyons slice into the land. On a geologic time scale mountains and mesas are as insubstantial as clouds; rivers are as vaporous as mist.

Cutler group — ?— *White Rim formation of Cutler group* — ?

270 MYA **265 MYA**

How is it that given the effectiveness of water and the passage of billions of years, jagged mountains still rise into the clouds and basins have yet to be filled with debris from the highlands?

Slow but inexorable movements within the crust of the earth keep thrusting some portions higher while plunging other portions lower. These movements prohibit erosion from reducing the face of the Earth to a near featureless surface. Here, mountains rise faster than erosion can annihilate them; there, erosion reduces a mountain chain to rubble as old mountain-building forces weaken or cease.

Revelations by Plate Tectonics

Crustal movement and deformation on a global scale is called "plate tectonics." Revelations made possible by the study of plate tectonics during the past few years have greatly enhanced our understanding of geologic processes. The story revealed by plate tectonics is fantastic, improbable, and indisputable.

Here are the basics: The crust of the Earth ranges from 5 to 40 miles thick. Its rocks are strong and resist being broken or deformed. Beneath this thin shell of ocean floor and continent lies the mantle, the greatest part of the mass of the Earth. Although most of the rock of the mantle is dense and strong, the rock of the upper mantle is relatively weak and plastic. Beneath the mantle lies the core, a region of very high density and pressure.

Inexorable and powerful currents within the upper mantle drive movements of the overlying crust. These movements have broken the brittle

◀ *Navajo Mountain reigns over a realm of naked rock and deep, sinuous canyons. But where did this isolated mountain come from? Look closely. Note that the rock layers tilt up the slopes of the mountain from all directions. Navajo Mountain belongs to a class of mountain termed "laccolithic." About 30 to 40 million years ago a mass of magma, molten rock, intruded from below. Normally such an event would signal the birth of a volcano. Instead, the magma flared out to the sides, muscling the sedimentary layers apart and creating a great blister of arching rock on the surface above. Laccolithic mountains are unusually common within the Colorado Plateau.*

? x x x x x x

240 MYA 235 MYA

◀ *Much of the Canyonlands rock column is visible from Elaterite Butte in Canyonlands National Park. In the extreme foreground, Shinarump Conglomerate (a member of the Chinle formation) caps red-sloped Moenkopi formation. More Chinle leads upslope to sheer cliffs of Wingate Sandstone capped with remnants of Kayenta Sandstone. White Rim Sandstone forms the edge of Horse Canyon beyond Elaterite Butte. The red Moenkopi lies above the White Rim while Organ Rock formation lies just below.*

Shinarump member of Chinle formation ⚌?⚌ *Mossback member of Chinle* ——— ⚌?⚌ *Upper Members*

——— TRIASSIC PERIOD ———→

230 MYA

No other rock layer within the Colorado Plateau can match the outrageous splashes of color within the Chinle formation. The rock is hardly rock at all—most of it is very soft and exceedingly crumbly. It originated from floodplain, pond, stream, and lake deposits. Volcanic ash, blown here from distant eruptions, also contributes to the great thickness of Chinle formation, 400 to 600 feet, in the Canyonlands National Park area. Petrified wood is common in some members of the Chinle formation.

crust into plates. Currently there are six large plates and several small ones. They make up the entire surface of the Earth, including the ocean floors, and move from one-quarter inch to five inches per year. Interactions between the migrating plates produce slow-motion collisions, sideswipes and rifts that drive geologic change — events large and small. Plates are not immortal. They may change shape or change location; they may split apart or join together.

The continents are composed of relatively low-density rock. They therefore "float" upon the heavier crustal rock, and their highest regions rise well above sea level. When plates move, so do the continents riding upon them. Hence, the continents are involved in the plate interactions. Continental collisions wrinkle the crust into new mountain ranges. The Appalachian Mountains in North America and the Atlas Mountains of Africa formed when two former continents collided and coalesced about 250 million years ago.

Oceans appear when continents pull apart along a plate boundary called a *spreading center.* About 200 million years ago the Atlantic Ocean began to open as an old continent split apart with a change in plate movements. Plate movements can cause the crust, continental or oceanic, to slip and grind along the crust of an adjacent plate. The San Andreas Fault is an example of such an interaction called a *transform fault.* When oceanic crust bends and dives beneath continental crust along a *subduction zone,* volcanoes often flicker to life near the edge of the continent.

Although it sounds like a nightmare planet

of Chinle formation →

225 MYA — 220 MYA

White water! When Lake Powell was near capacity, the San Juan River dumped huge quantities of mud and sand into the slack water. Later, when the lake lost volume, a broad mud flat was exposed. The San Juan River pushed out across the soft sediments and entrenched as the lake retreated. This new silt-bordered "canyon" was superimposed on the old landscape buried in mud. Here the river tumbles over a low cliff, still searching for its former bed that lies lost in a sandbox of lake sediments.

Seeps at the base of a cliff may eat away support, or an underlying softer rock may crumble, or flash floods may batter away the foot of a cliff. For whatever reason, the Wingate Sandstone above has fallen away leaving a stable arch form in compression to support the rock above.

Rainbow Bridge stands 290 feet high and spans 275 feet. Bridge Creek has a history of discarding old channels for new ones as it occasionally shortcuts its meanders. One of these breakthroughs left the cliff above standing firm. Frost and heat, flood and drought further shaped the opening into the form of a rainbow.

Upper Members of Chinle formation

TRIASSIC PERIOD →

215 MYA

of a sinister universe, this is our own Planet Earth. Fortunately, the vast majority of these titanic events are imperceptible to us transients.

Geologic Subdivisons

Geologists have been able to subdivide North America, the continental highland of the North American Plate, into a few provinces that have a common history or structure. The Lake Powell region is at the heart of the Colorado Plateau Province. It is centered near the confluence of the Colorado and San Juan rivers, and includes the southeast half of Utah, northern Arizona, and small portions of New Mexico and Colorado.

Geologically the Colorado Plateau may be characterized as relatively high in elevation, composed primarily of flat-lying sedimentary rock layers cut into by a rather large river, and bounded by provinces of highly broken or folded strata.

The area we know as the Colorado Plateau was once part of the stable interior of the continent. (Continental interiors tend to be quiescent while continental margins are involved in plate boundary adventures!) This interior province included layer upon layer of flat-lying sediments in a low-elevation position. But then about 85 million years ago a sequence of events began that isolated and transformed a portion of the province. In the course of 20 or 30 million years, the southwestern lobe of the stable interior was cut off by events of crustal deformation. To the northeast the Rocky Mountains formed as the crust folded, fractured, and warped into highlands. To the southwest the Basin and Range Province rose to block

Wingate formation

JURASSIC PERIOD →

210 MYA

205 MYA

▲ *Buckskin Gulch rips a zigzag trench across the Paria Plateau. For 12 miles Buckskin Gulch averages 15 feet wide! Backpackers sneak through its gloomy corridors toward a meeting with the Paria River. But chilling pools, knee-deep mud, occasional debris dams, and flash flood threats dictate that they be familiar with the perverse habits of slot canyons. Note that Buckskin Gulch sporadically follows a system of joints visible on the plateau surface.*

Kayenta formation ?*Navajo formation*

JURASSIC PERIOD →

200 MYA

the exit of rivers draining the plateau area. The plateau became a basin receiving sediments eroded from the surrounding highlands.

THE COLORADO RIVER

Periods of uplift followed, but until about 5.5 million years ago when the Gulf of California opened along the San Andreas Fault, the Colorado River did not exit the plateau to the southwest. There were long periods of ponding when there were no outlets to the sea. An ancestral Colorado may have exited to the northwest of the Grand Canyon area shortly before the formation of the Gulf of California.

Starting about 5.5 million years ago, the Colorado Plateau and the surrounding areas were rapidly uplifted about 3,000 feet. The Colorado River, fortified by the steepened gradient and having exploited the shortcut exit to the sea via the Gulf of California, developed into the powerful river we see today. The canyons of the Colorado were cut with great speed by the enlivened river.

With mountains pushing up here and there, with the surface of the earth warping, undulating, and breaking, with former outlets obstructed, it is difficult for a river to prosper. The Colorado may well be of different ages along its length — older along its upper reaches, younger along its lower end.

So, how old is the Colorado River? Its age is difficult to assign — the river has evolved. At what point in time was it so different from today's river that it cannot honestly be labeled the Colorado River? How old? Five and a half (5.5) million years if you're persnickety, older if you're not.

Advancing sand dunes may have once buried a section of Antelope Canyon. When wetter weather returned, a lake formed behind the sand-dune dam. Eventually the rising water overtopped the dam and flooded downslope near, but off to the side of, the original canyon. A new canyon segment took shape as every succeeding storm deepened and developed it further. The canyon's walls retain the shape of the fast-moving water because weathering has not yet had time to soften evidence of the rapid and geologically recent canyon-cutting. ▶

Rocks — Their Creation and Destiny

To many visitors of the Lake Powell region the rock layers become treasured friends recognizable in different guises and under odd conditions. Each strata's strengths and foibles, imperfections and quirks have been revealed by encounters in diverse locations spread over years of experience. Occasionally one of these rock buddies will pull a fast one, but the surprise keeps the relationship stimulating and fresh!

Types of Rocks

Most of the rock layers within the Colorado Plateau are blankets of sediments that accumulated on land or in water. Over periods of millions of years these layers of debris compacted and hardened into strata called *sedimentary* rocks.

Two other types of rock can be found on the Earth's surface. Both exist in the Lake Powell region, but are far less common. *Igneous* rocks, those rocks originally deposited in a molten state, can be found in some Lake Powell region mountains, or as lava flows or lava boulders. *Metamorphic* rocks, the third type of rock, were originally sedimentary or igneous but have been so altered by heat and/or pressure that they possess characteristics very different from their initial forms. In the Lake Powell region metamorphic rocks are most likely to be found as cobbles carried here from distant sources by the Colorado River and its tributaries.

◀ *Hourglass Rock near Navajo Mountain stands alone amid a tangled landscape of domes, hollows, cliffs, canyons, and knobs. Most of the exposed rock is Navajo Sandstone, massive and pleasing to the eye. The grain-size of the sand, the amount and type of cement that binds the sand grains together, the local tilt, and the abundance of joints play interesting variations on a theme of mesa and canyon, pedestal and slot.*

Navajo formation ——— ?— *x* *x*

——— JURASSIC PERIOD ———→

185 MYA

What are these hard rocks, quartzites, lavas, and limestones doing in this land of soft sandstones and shales? Their rounded shapes and bench-top positions are major clues. The cobbles were carried here from the mountains of Colorado and Wyoming by the Colorado River. This group is perched far above the present river level. They were dropped here when the riverbed was 500 feet higher than today, probably during a flood many hundreds of thousands of years ago.

▲ *Monument Valley's Organ Rock Shale is soft and susceptible. It crumbles beneath the cliff-forming DeChelly Sandstone above it. Inevitably the DeChelly, cantilevered out over the feeble shale, sheers and falls. When the dust settles, the countless shattered boulders, cracked rocks, chips and flakes offer a vastly increased surface area, ripe for attack by water and frost.*

The Origin of Cross-bedding in Wind-deposited Sandstones

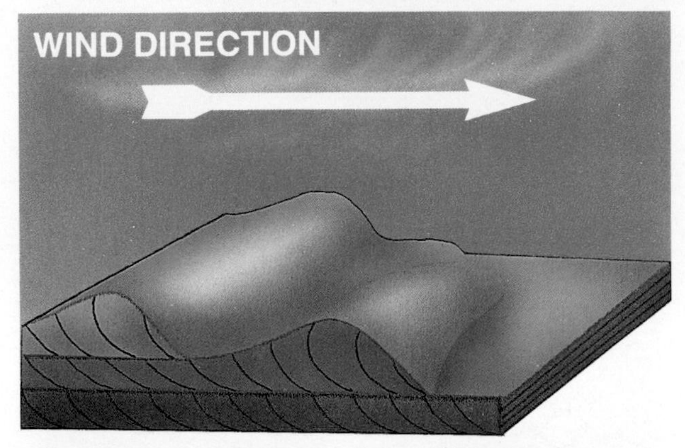

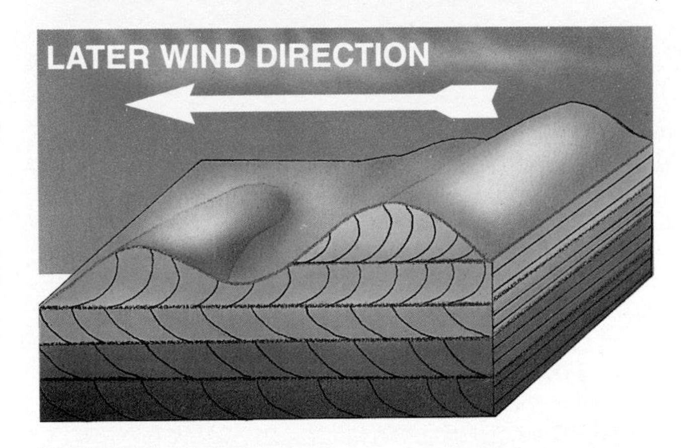

Trench a modern sand dune in Monument Valley and you will discover the same complex bedding patterns displayed in the cliffs and domes of eolian (wind-deposited) DeChelly, Wingate, and Navajo sandstones. The abundance of sand, strength of the wind and other factors determine the external form and internal structure of a dune. As a barchan-type dune advances, for example, the gently sloping upwind face erodes and the steeply sloping downwind face builds. If the hungry upwind slope fails to eat away the dune's full height, a low trail of tilting downwind beds is left behind. When the winds shift, the dunes reorganize. Cross-bedding patterns record the passage of succeeding dunes and wind-shift reorganizations. The most fanciful patterns are often produced by the simple interaction of undulating erosion surfaces and the ordinary patterns of deposition.

x x x x x

JURASSIC PERIOD →

170 MYA

The Entrada Sandstone at Goblin Valley State Reserve in Utah is eroded into a land of wacky shapes. The rock is soft and easily eroded. The goblins range in size from one to thirty or forty feet in height. The laccolithic Henry Mountains loom on the horizon.

Sedimentary layers are records of prior environments or conditions that existed before the current environment developed. Seabed deposits often turn to limestone, sand deposits become sandstone, and mud deposits become shale.

If a layer of limestone is revealed in a canyon wall, the ocean that produced the limestone did not once fill the canyon, it *predated* the canyon. Do not indulge the canyons and slight the clock! Canyons are ephemeral, rock layers are long-lasting, but time is eternal — at least in our universe.

Of course, deposition cannot proceed indefinitely. Crustal movements will interrupt the process. Layers of material that have turned to rock will likely be uplifted, exhumed, and exposed to weathering and erosion. Water, with some help from wind and glaciers, will carry them away, particle by particle, to a new basin of sedimentation.

Thus, no single location retains a perfect record of all the environments that once existed there. Portions of the record are missing because some conditions dictate a hiatus in deposition. Or a period of erosion may have removed layers or portions of layers. The resultant gap in the stratigraphic sequence is called an *unconformity.* Unconformities represent a substantial break in the record, and make more difficult the deciphering of the local geologic story.

Sedimentary rock layers exhibit a number of

Overleaf: Sand, once locked in the dunes of a great desert, is set free again by time and the agents of erosion.

Dewey Bridge member of Entrada formation (Carmel formation to south) — *Slick Rock member*

165 MYA **160 MYA**

Slick Rock member of Entrada =?= *Moab member of Entrada formation (May be equivalent t*
JURASSIC PERIOD
155 MYA

l or part of Curtis and Summerville formations)
?
Morrison formation
?
150 MYA
145 MYA

◀ *What's this, a ledge of limestone within the Navajo Sandstone? Sure enough! These thin lenses of sandy, freshwater limestone are believed to have come from small lakes or oases cradled between the dunes in the Navajo desert. The ledges extend only a few hundred feet before pinching out.*

The spires and columns of Bryce Canyon are composed of Claron or Wasatch formation. The rock is quite young, only about 50 million years old, and is a poor grade of limestone. The pinnacles and towers are as fragile as they look. A number of named features have collapsed into rubble during the past few decades. Geologists estimate the rim of Bryce Canyon is receding at the rate of about one foot per 60 years! ▶

interesting characteristics: Their extent varies widely from a few square miles to half a continent. They range in thickness from a few feet to several thousand feet. The boundaries of these layers thin to disappearance, grade into or interfinger with adjacent deposits. These rock-layer boundaries or transitions are records of environmental boundaries or transitions—river deltas to oceans, swamps to highlands—just as we see today.

The Stratigraphic Name Game

The naming of rock layers follows a strict protocol occasionally violated in devilish ways. The stratigraphic name game is played in this way: The basic unit of stratigraphy is the *formation,* a layer or group of layers.

A formation's name usually comes from the geographic locality where it is best exposed, or from where it was first studied in detail. When appropriate, formations are subdivided into *members* with names of their own, such as the Petrified Forest member of the Chinle Shale.

Several formations are sometimes gathered into larger units called *groups.* The name game exasperates new players when a rock unit is inadvertently tagged with two or more names, its continuity initially being unrecognized. The game further provokes newcomers when a member in one location is christened a full-fledged formation in another location. Also, the names are occasionally changed as more is learned about the rock units and their interrelationships.

Once they have been deposited, strata are commonly subjected to stresses that disturb their original flat-lying character.

Despite a history of great uplift, the rock strata of the Colorado Plateau have generally retained their flat-lying demeanor, a fundamental characteristic of the province. In those locations where the strata have been flexed or broken, the feature is usually vivid and distinctive, unspoiled

Thousands or tens of thousands of years ago resistant conglomerate boulders tumbled from a cliff to catch on a slope of soft siltstone. Today those same boulders teeter on siltstone pedestals because the surrounding slope has melted away in the rains. Only slender remnants of the old slope remain where they huddle under the safety of the conglomerate boulder umbrellas.

The Waterpocket Fold of Capitol Reef National Park is believed to be about 65 million years old.

135 MYA 130 MYA

Anatomy of a Sandstone Dome

Joints and Fractures: *Flexing of the rock mass over millions of years has cracked the rock strata. These ancient joints are zones of weakness that invite the attack of water. Where the fractures occur in groups, as is common, the slabs between the joints crumble and fall away.*

Color: *Most of the Navajo Sandstone is composed of clear quartz grains cemented together by minerals. The minerals, often deposited by groundwater long after the accumulation of the sand, greatly influence rock color. Iron oxide, for instance, creates much of the reds, oranges, and tans typical of Navajo Sandstone.*

Cross Stratification: *(Also called cross-bedding.) As sand accumulated 190 million years ago, a detailed but incomplete record was made of shifting wind directions and dune movement. The tilted beds usually dip downwind.*

Mass Wasting: *Often, gravity without the aid of water or wind, moves rock fragments. Here, slabs of projecting erosion-resistant ledges have yielded to gravity and have broken off from the parent rock a few feet above.*

Petrified Sand Dunes: *Navajo Sandstone is often characterized as a mass of "petrified sand dunes." Indeed, the Navajo Sandstone is composed primarily of wind-blown sand from a vast ancient desert. However, today's dome-and-hollow, undulating landscape does not re-create the terrain of 190 million years ago.The shape of today's slickrock topography, as with any landscape, is determined by a number of interdependent factors including the nature of the rock, location and abundance of joints, local and regional drainage patterns, vertical relief, and climate.*

x *x* *x* *x*

CRETACEOUS PERIOD →

125 MYA

x=Significant Periods of Erosion or Non-Deposition

by complex or severe deformation. Such unmistakable folds and faults are a joy to behold, especially when only sparse desert vegetation obscures the view.

Strata-distorting Features

There are several different varieties of strata-distorting features, and every one of them perpetrates clever tricks in an otherwise flat-featured tabular landscape.

Faults, linear breaks in bedrock, in the Lake Powell region are fairly common, but tend to be of modest displacement. (Other regions of the Colorado Plateau exhibit faults of great displacement.) They are typically inactive. Movement along these faults has been primarily vertical. Faults are sometimes manifested as escarpments or cliff lines.

But, do not be misled. Most cliffs in the region have formed as the result of swift downcutting of the Colorado River and its tributaries and/or the effect of monoclines.

Joints are breaks in bedrock that show no movement one side relative to the other. Canyons and side canyons often develop along joints and faults because the fractured rock offers zones of weakness that can be exploited by erosion.

In Canyonlands National Park a series of parallel faults have developed, perhaps as underlying deposits of salt and gypsum have migrated or dissolved away. In some places the roof rock has collapsed along these fault lines to create a number of parallel, shallow, sheer-walled canyons often 300 feet deep. A canyon formed in such a manner is termed a *graben,* a German word meaning ditch or grave. The adjoining blocks that remain high relative to the grabens are known as *horsts.*

It is not difficult to spot the mysterious "lanes" of the Needles District of Canyonlands National Park. Two sets of parallel fractures, oriented roughly 90 degrees to each other, crisscross the Cedar Mesa Sandstone benchlands. About 500 feet of faulting has displaced the blocks between some of the northwest-southeast trending fractures. The horsts (uplifted blocks) form the walls of the lanes while the grabens (down-dropped blocks) underlie the lanes. Like all salt deposits, the salt beds within the older and deeper Paradox formation, under pressure from the overlying rock, are plastic. Migration and dissolution of the Paradox salt deposits are believed to have caused the faulting.

x x x x

120 MYA **115 MYA**

Square Butte on the Navajo Indian Reservation is an example of a stage in the evolution of topography. Softer rock at the base of a mesa (Carmel formation, in this case) disintegrates and undermines a more resistant rock (Entrada formation) above. Gradually the mesa (wider than it is tall) shrinks to the proportions of a butte (taller than it is wide) that eventually erodes to a spire that ultimately collapses into a mound.

In many locations stresses exerted laterally across the crust of the Earth have folded rather than broken the strata. A trough-shaped downfold is called a *syncline.* An arch-shaped upfold is called an *anticline.* Synclines and anticlines are often found in series such as along the San Juan River downstream from Bluff, Utah. Another type of fold, called a *monocline,* is half a syncline or anticline. That is, one nearly flat-lying side is high relative to the other nearly flat-lying side with a steep slope linking them. Monoclines are unusually common in the Colorado Plateau.

Examples of monoclines in the Lake Powell region are the Echo Cliffs monocline crossing the Colorado River at the foot of Glen Canyon and the Waterpocket Fold in Capitol Reef National Park. Another unusual feature of the Colorado Plateau is its abundance of *laccolithic* mountains. In the Lake Powell region the Henry Mountains, the Abajo Mountains, and Navajo Mountain are all laccolithic. Here the sedimentary layers were blistered upward as molten rock invaded from below. Normally a volcano would soon erupt. But here the lava failed to force its way to the surface. The molten body cooled and solidified. Recent erosion has in some mountains exposed the core of igneous rock surrounded by concentric rings of tilted sedimentary beds.

The flat-lying strata interrupted here and there by folds and faults, and revealed by a vigorous river have made the Lake Powell region an outdoor textbook of elegant form and structure.

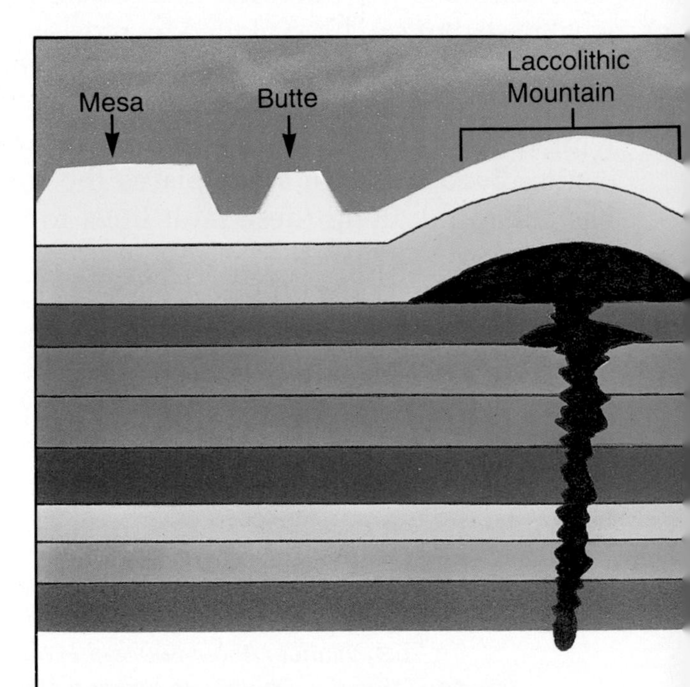

The Colorado Plateau is a physiographic province that includes parts of Arizona, Utah, Colorado, and New Mexico. Bare rock shows through a thin veneer of soil and plant life. Most importantly, the Colorado

CRETACEOUS PERIOD →

110 MYA

The Brain Rocks: Polygonal jointing sometimes dominates the surface of Navajo Sandstone. The joint patterns may result from the expansion and contraction of the rock near its surface where temperature fluctuations are most extreme. Polygonal jointing is often associated with areas where the cross-bedding has been disturbed by slumping shortly after deposition. The slope of a monocline lies in the distance beyond the Brain Rocks.

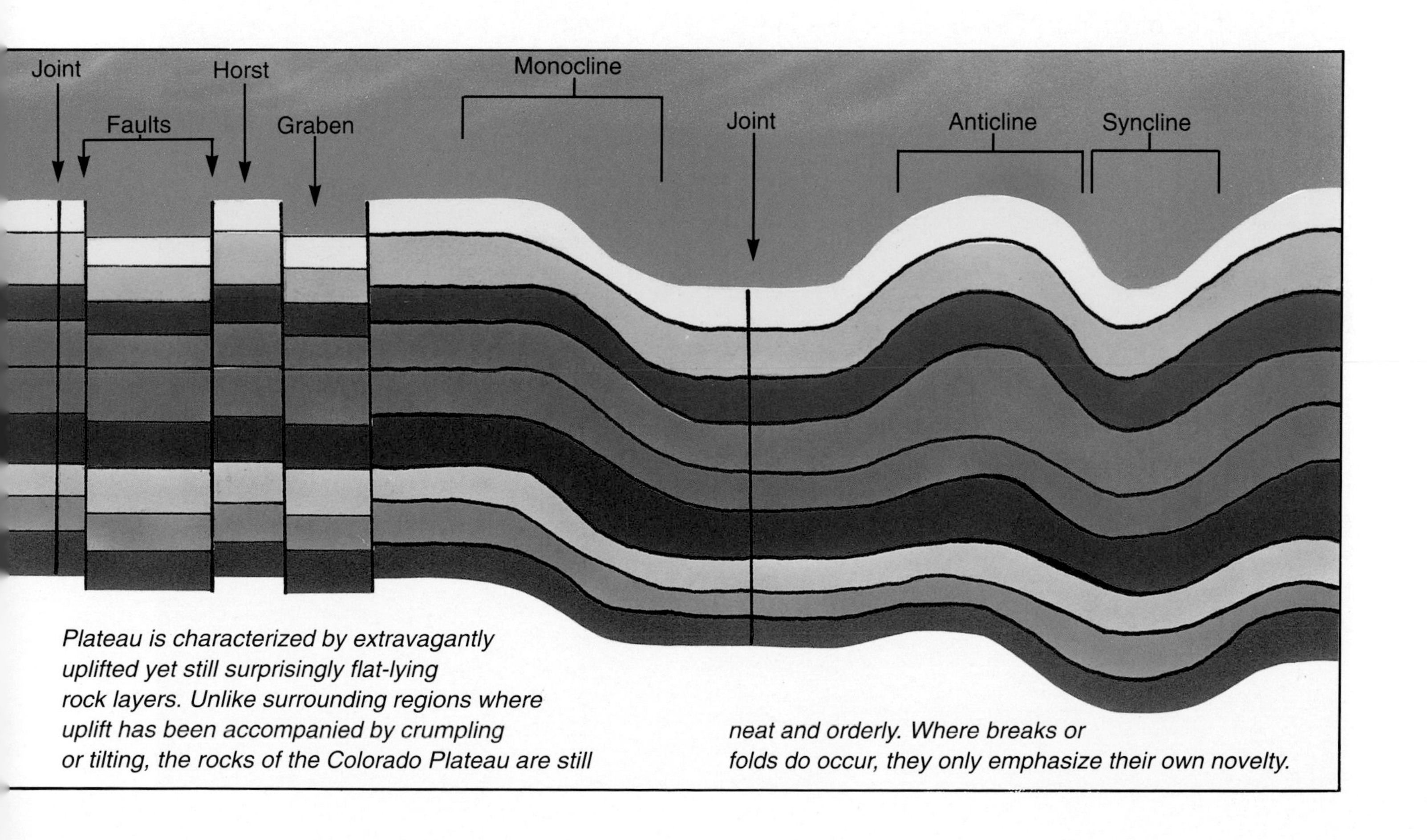

Plateau is characterized by extravagantly uplifted yet still surprisingly flat-lying rock layers. Unlike surrounding regions where uplift has been accompanied by crumpling or tilting, the rocks of the Colorado Plateau are still neat and orderly. Where breaks or folds do occur, they only emphasize their own novelty.

x x x x x

105 MYA **100 MYA**

Sandstone Fantasy

Sometimes it seems as if every square foot of the sandstone surface is of mysterious origin. What's happened here? Was it swirling water, unusual depositional circumstances, fracturing and cementation, differential erosion, slumping of water-saturated dune deposits, or a combination of causes that produced THAT fantastic form?

–?– *Dakota formation* ——————— –?– *x* *x* –?– *Mancos*

——— CRETACEOUS PERIOD ———→

95 MYA

formation (Tropic formation to southwest) ⟶

90 MYA

85 MYA

Water and the Forces of Erosion

Our journey through geologic time now concludes with a flourish as the finest touches are placed upon the landscape. After having found evidence of cycle upon cycle of erosion and deposition, the pendulum of change sweeps into a period of erosion once more. It is an erosional period of astonishing proportions creating an intricate and beautiful landform.

Do Not Be Deceived

There is a pitfall here, however. We admirers of the Lake Powell region may be so intrigued by the grandeur of this landscape that our perception of geologic history is distorted. Do not be deceived. The present dominance of erosional features — the canyons, mesas, buttes, alcoves, pedestal rocks, and sandstone slots — represent a momentary puff in the swirling winds of time and change.

Consider the Navajo Sandstone, the formation into which Glen Canyon Dam is keyed. It contains the remains of a vast sandy desert that lay here almost 200 million years ago. Yet it was succeeded by tidal flats, lake and stream deposits and a number of marine sediments, all of which existed long before today's canyons were cut. Furthermore, the Navajo Sandstone desert was preceded by dozens of earlier environments, many of them represented by the rock formations found beneath the Navajo Sandstone today.

◀ *A plume of water leaps from the rim of Glen Canyon after a summer downpour. But the waterfall is not composed solely of water—sand and silt are also swept along in the current. And when the rains are torrential the resulting floods heave cobbles and boulders over the cliffs, too. Remember to wear a hard hat while bathing under this shower!*

--- —"Isolation" of Colorado Plateau due to uplift of surrounding areas (Laramide Orogeny) —

—*Mancos formation* — ≡?≡ *x* *x* ≡?≡ *Straight Cliffs formation*

— CRETACEOUS PERIOD →

80 MYA

◀ *Look carefully into the depths of Cataract Canyon. . . and watch your step! Amazingly, here, just below the confluence of the Green, the Colorado flows along the length of an anticline, a stratigraphic ridge! It is clear that the easily deformed salt deposits of the Paradox formation have had a hand in this peculiar circumstance, and it seems likely that the anticline existed before the Colorado River arrived. But WHY the Colorado followed the strike of the Meander anticline remains a mystery.*

—?— *Wahweap formation* —?—

75 MYA 70 MYA

Nevertheless, it is erosion, the wearing away of rock, that pervades our view. The first step in the ruin of rock is weathering, the decay of rock subjected to the atmospheric elements of moisture, air, and organic matter. It occurs within a zone extending to whatever depth air and water may penetrate, often tens or hundreds of feet. Weathering includes both chemical decomposition and mechanical disintegration.

Erosion transports away the weathered fragments. More than with any other geological process, erosion is observable, approachable, and understandable. It is in action now, this very moment. Pick up a rock. Now kiss it goodbye! Simply by being on the surface and small enough to heft by hand, it is on the edge of oblivion. It will soon be dismembered and its particles rudely escorted to a basin repository.

During a thundering flash flood the cobbles and boulders that lie on a canyon floor will begin to tumble downstream. The bedrock floor of this canyon may be five, ten, or more feet below the level hikers walk upon in dry weather.

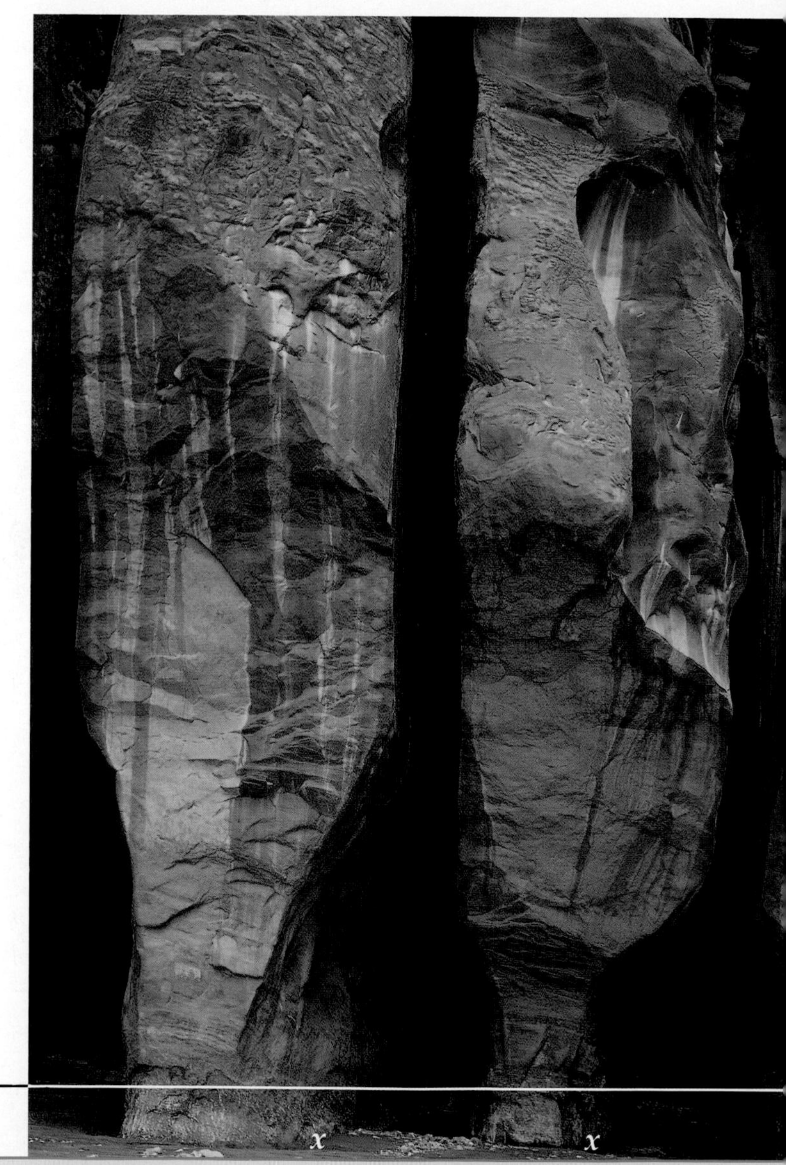

Joints in the Navajo Sandstone often cross the Paria River. When the Paria floods the swirling, gritty water pelts these joints, zones of weakness, with hard stones transported from upstream, widening the cracks to chambers.

Kaiparowits formation ?

CRETACEOUS PERIOD → TERTIARY PERIOD — PALEOCENE EPOCH →

65 MYA

CENOZOIC ERA →

Follow That Rock!

Until recently that rock was part of a rock formation. Hundreds and thousands of feet of younger rocks lying above protected it from erosion. But the uplift of the Colorado Plateau, mostly within the last 5.5 million years, has rendered it vulnerable. Bit by bit the overlying strata were stripped away. How and by what agent was this monstrous task accomplished?

Water mostly. Just water and time, those deceptively simple powers. Water dissolves the cement that binds the rock particles. Near the surface of the Earth water can invade tiny fractures. When temperatures drop below freezing, frost wedging splits the rock. Or perhaps the rock is betrayed by softer layers beneath — it is undermined and eventually shears from the cliff, a victim of *mass-wasting.*

Concretions range in size from pinheads to cannonballs and occur occasionally within the Navajo Sandstone. They are harder than the encasing rock, and often erode out to roll downhill and gather in nearby basins. Concretions result from the local concentration of cementing agents present in the groundwaters that surround the sand grains after deposition. When they lie loose upon the ground, concretions are sometimes called "Moki marbles."

This is not a view of the moon's surface. It is a couple of square yards of sandstone bedrock embedded with an unusual type of concretion layered like an onion. When viewed in erosional cross section the concretions look like miniature craters.

"Isolation" of Colorado Plateau due to uplift of surrounding areas (Laramide Orogeny)

x x x x x x

EOCENE EPOCH →

60 MYA **55 MYA**

▲ *Gunsight Butte is reflected in the serene surface of Lake Powell while Navajo Mountain looms on the distant horizon. At the foot of Gunsight Butte, broken rock has collected in piles resting against the cliffs. Each pile is called a "talus." A talus is an accumulation of weathered rock fallen from the cliffs above. Its constituent rocks rest at the steepest angle at which the slope remains stable, usually in the range of 33 to 37 degrees.*

Wasatch (Claron) formation ? x x

EOCENE EPOCH →

50 MYA

Severed from its parent rock layer, the pace of decay accelerates. The rock now lies at the top of a *talus slope,* a collection of rock fragments, large and small, derived from strata above. Gravity tugs at the rock, doggedly urging it downslope. In fits and starts it does so. Imperceptibly it slides or slips aided by rain and frost heaving. Occasionally it tumbles; corners are knocked off. Weathering continues to attack all exposed surfaces.

Over a period of thousands of years the rock nears the foot of the slope. Infrequent but powerful flash floods eat away at the slabs and shards. Finally the rock is snatched by a frothing brown flood shortly after a summer downpour. The flood ruthlessly rolls the rock for a mile before it catches in a crack. Here it stays for another five hundred years before being dislodged by a more vigorous flood. The rock tumbles and bounces and slides within a gritty soup of boulders, pebbles, logs, and twigs. It stops again as the flood subsides, this time only a mile from the Colorado River. Flash-flood battering has quartered its diameter.

In a thousand years more it moves to within a hundred yards of the Colorado. Here it rests for three centuries. Finally, with the side canyon commencing to flood once more, the rock shifts,

HEART OF THE COLORADO PLATEAU GALLERY

Each rock layer is the debris of some past environment. Like today's environments, those of the past did not often possess sharp-edged boundaries. Accordingly, rock units also lack definitive margins. They gradually pinch out, they interfinger, they blend one into another. Thus no single rock column diagram can accurately portray the entire Colorado Plateau. Yet the stratigraphy of the heart of the plateau is consistent enough throughout to form a generalized schematic rock column. On display are many of the prominent rock units of the Heart of the Colorado Plateau, their relative ages and some of their important features. The exposed rock sequence of the Colorado Plateau is a geologic three-dimensional puzzle, just easy enough to be alluring—just hard enough to be intriguing.

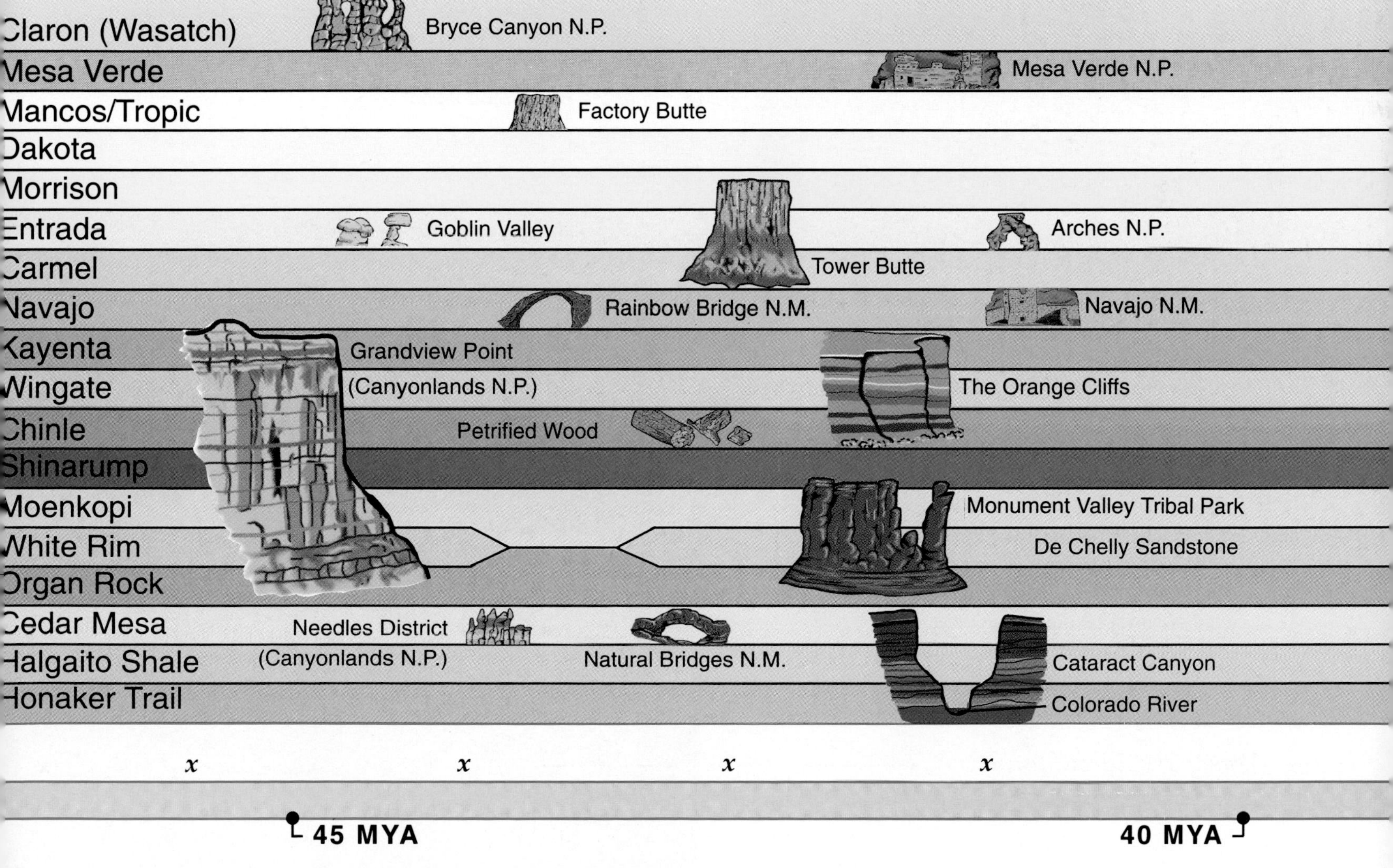

▲ Fingers of desert varnish reach down Navajo Sandstone alcove walls toward Keet Seel Anasazi Indian ruin. The formation of desert varnish, or patina, is not fully understood but the thin, hard film is composed primarily of iron and manganese oxides leached from the rock, and dust blown on the wet surfaces during rainfalls.

▲ Navajo Sandstone often breaks in a distinctive manner called "conchoidal fracture." Cliff faces are often marked by the sinuous, concentric ridges characteristic of conchoidal fracture.

Look at those fins! Course-grained rocks such as this Navajo Sandstone typically possess widely spaced joint patterns while the joint systems of fine-grained rocks, such as shale, are closely spaced, minute, and far more numerous. This choppy sea of rock fins, the solid rock blocks between joints, lies ▼ near the Henry Mountains.

Intermittent laccolithic intrusions

x x x x

EOCENE EPOCH | OLIGOCENE EPOCH

35 MYA

◀ *The Slick Rock member is relatively permeable, the Dewey Bridge member beneath it is relatively impermeable. Cliff face seeps develop where the water fails to penetrate the Dewey Bridge. The seeps attack the rock cement, the rock disintegrates, and alcoves form where the cliff is undermined.*

hesitates, then suddenly bounds for the master river. Tens of tons of cobbles and boulders bury it at the confluence. A riffle, a small rapid, forms where the Colorado encounters the fresh collection of flood-borne boulders. For a hundred years our rock will remain in the bowels of the riffle. Gradually, with each spring flood, the riffle moderates as the boulders are rasped and agitated by the silty torrent. When the rock breaks free it joins the irresistible downstream migration.

Within a few decades it has been pulverized by the river and its hitchhiking load of rock and sand. Within a few more tens of years nearly every particle has been transported to the dark and distant ocean.

Here the cycle begins anew. The rock bits will be incorporated into layers of debris. In millions of years it will recycle into rock. In half a billion years it may emerge as a component of a limestone crag on a jagged mountain, or a sandstone layer underlying a plain. All it takes is time . . . just time.

Has there been displacement along a fault line here? ▶ *No. The left wall is coincident with an ancient, hard-surfaced joint face. The right wall is hollowed from a solid block of rock between joints. Blue sky illuminates the dark wall—reflected sunlight the right wall.*

——— Intermittent laccolithic intrusions ——— - - -

x *x* *x* *x* *x*

30 MYA **25 MYA**

◁ *Flash flood! A tongue of surging, brush-laden water, edged in white foam, races down Cliff Canyon. A "freight train" roar precedes and warns of the approaching torrent. Narrow canyons with sheer walls keep hikers alert in rainy weather.*

x x x x x

MIOCENE EPOCH ⟶

20 MYA

THE IMPORTANCE OF FLOODS

Note the importance of the floods. Floods must cart away all the rock quarried in all the tributaries of the canyon. The larger the flood the faster the flow; the faster the flow the better equipped it becomes to move large blocks and large volumes.

Once, as a member of a small group of backpackers, I camped in a tributary canyon of Lake Powell. A downpour forced us to take shelter in an alcove overlooking the confluence of Cliff and Forbidding canyons. We soon heard a deep and ominous rumble. From around the bend gushed a probing tongue of angry frothing water.

As the rumble escalated into a roar, the flood grew in width and depth. We heard cobbles and boulders clattering along the streambed. We watched as the flood surrounded then swallowed a sandstone boulder six feet in diameter. From the confluence the rising flood turned both downstream and upstream in Forbidding Canyon. I had never imagined that a flash flood could flow upstream, even if for only a hundred yards.

For half an hour it raged. In four hours the flood shrank to a stride wide. By the next morning it was a trickle. We estimated the maximum flow to be about 1,000 cubic feet per second, about one-tenth of a typical Colorado River flow.

Much more rare are debris flows, high density mudflows, often of monstrous proportions. A thick mixture of water, rock, and soil the consistency of wet concrete makes a flash flood look trivial in comparison.

◀ The flash flood recedes. A few minutes after the first wave of mud-laden water rounded the bend, the depth of the flood topped six feet. Now, an hour later, maximum water depth has fallen to about two feet. Yet cobbles continue to clatter down the bed of the tumbling stream. In another three hours the stream will have nearly returned to its pre-flood flow, but with a rearranged setting of cobbles and boulders.

▲ Tinajas (water pockets) dot the slickrock near Moab, Utah. The natural basins catch rainwater which deepens the basins by dissolution.

▲ A tale of alternating flood and desiccation is revealed in a cliff of Triassic mudstone. The story is complex but notably the white knobs are the more resistant sandy fillings between slightly older mud cracks.

--- ——— Highlands west and southwest of Colorado Plateau subjected to rifting ——— ---

x x x x x

15 MYA **10 MYA**

▲ *Near Lake Powell, backpackers explore a rugged Navajo Sandstone landscape seamed with joints.*

—— Intermittent laccolithic intrusions and basaltic volcanism ——

Colorado Plateau and adjacent areas uplifted about 3000 ft.

Colorado River drainage begins canyon cutting.

x *x* *x* *x* *x*

—MIOCENE EPOCH—→ | PLIOCENE EPOCH →

5 MYA

Canyons Are Just Narrow Valleys

Here is a final irony — canyons are just narrow valleys with high, steep walls. The recent rapid downcutting of the Colorado River has produced steep-sided valleys whose "banks" have had little time to recede from the roaring river. If these valleys were bigger still, tens of miles wide with gently sloping walls and rims obscured by distance, our perception of their size would be diminished rather than enhanced.

The modest volume of rock removed, the sheer cliffs, the proximity of the rims, all aided or made possible by rapid downcutting have upgraded the valleys to canyons. Canyons look big because they are surprisingly small relative to their depth!

SUGGESTED READING

Baars, Donald L. *Canyonlands Country.* Lawrence, Kansas: Canon Publishers, 1989.

Chronic, Halka. *Pages of Stone No. 4, Grand Canyon and the Plateau Country.* Seattle, Washington: The Mountaineers, 1988.

Hartmann, William K. *The History of Earth.* New York: Workman Publishing, 1991.

Johnson, David W. *Canyonlands: The Story Behind the Scenery.* Las Vegas, Nevada: KC Publications, Inc., 1989.

McPhee, John. *Basin and Range.* New York: Farrar Straus Giroux, 1980.

Rahm, David. *Reading the Rocks.* San Francisco: Sierra Club, 1974.

Shelton, John S. *Geology Illustrated.* San Francisco and London: W. H. Freeman and Company, 1966.

Skinner, Brian J. and Stephen C. Porter. *The Dynamic Earth.* New York: John Wiley & Sons, Inc., 1989.

Tomorrow and the Time Line

The very landscapes that we equate with "timelessness" are those that are among the youngest. The canyons were excavated only yesterday.

Geologists continue their quest to pinpoint dates. But the precise times of sedimentary rock deposition are often shadowy and the true times of deformation are sometimes sketchy. The very fact that the sediments of the Heart of the Colorado Plateau lie flat complicates their interpretation. Around the campfire field geologists still argue about the exact age of some rock units. Only further snooping will reduce the disagreements.

We can be sure that for about 250 million years a thick accumulation of sand, silt, pebbles, and lime gathered on top of a pre-existing set of rocks and ancient faults. We can be sure that these sediments became the raw material from which the Colorado River and its tributaries shaped the spires and cut the clefts of the Heart of the Colorado Plateau. We can see that the result is a splendid exhibit of geological science presented in the form of art.

Upon these rock strata recently carved into a thousand curious shapes, few younger units were ever created—plateau uplift commenced to curtail the supply of particles and pebbles from which strata are fashioned. Weathering and erosion took charge and they continue their sovereignty to this day.

Still entombed below today's landscape are older rocks constructed when Earth was a harsher, more dangerous locale. We know less about these rocks and their periods of creation because the evidence has gone stale or has been trampled by later events. Some time in the future these older strata may be laid bare and engineered into arches and alcoves along the Colorado River or an aggressive heir.

Mountain-building commonly persists for tens of millions of years. Many depositional environments such as coastal plains, deserts, and deltas carry on for millions to tens of millions of years. Mighty mountain systems, if they're lucky, survive several hundred million years. Most rivers evolve and mutate, and even prey on the weaknesses of lesser rivers until they are unrecognizable within millions to tens of millions of years.

Thus the time line inches into the future and today is NOT the end of the line!

QUARTERNARY PERIOD
x Holocene Epoch
PLEISTOCENE EPOCH
PRESENT DAY

It's a blistering July afternoon near Lake Powell. Dinosaur Rock, a stone caricature of a brontosaurus, surveys the scene from a hilltop lookout. "Dino" has been quietly observing from this outpost for a long time—for centuries, most likely. But today is different. As a tour boat rumbles by, Dino trembles, keels over, and gracelessly crashes into extinction. Too much sun for the old boy. Too much cold, wind, rain, and snow.

Landforms. They evolve. They take their turn in the sun and then surrender to upstart landforms. Their stories are told in the unmasked strata of the Heart of the Colorado Plateau. But not far away, at the foot of Glen Canyon, they nearly vanish. The Organ Rock formation becomes the Hermit Shale. The White Rim becomes the Toroweap. The Grand Canyon begins. But that's another story—another fantastic landform!

◁ *Diagenetic Arch frames Boundary Butte near Lake Powell. Its "rolled" shape stems from the locally contorted beds of Navajo Sandstone.*

A Magnificent Landscape

The Lake Powell region, the heart of the Colorado Plateau, is a staggeringly magnificent landscape. A combination of unlikely circumstances have conspired to create a visually pleasing composition. Like a Bach fugue, the canyonlands are logical, coherent, and satisfying to the senses. Knowledge of the geologic *story behind the scenery* serves to increase our delight in this symphony of stone.

When summer comes this snow melts and tumbles down the mountain sides in millions of cascades. A million cascade brooks unite to form a thousand torrent creeks; a thousand torrent creeks unite to form half a hundred rivers beset with cataracts; half a hundred roaring rivers unite to form the Colorado, which flows, a mad, turbid stream, into the Gulf of California.

— John Wesley Powell.

Published by KC Publications, 3245 E. Patrick Ln., Suite A, Las Vegas, NV 89120.

Inside back cover: *Thin ribs of resistant sandstone protrude from the midsection of a wind-blasted goblin.*

Back cover: *Ripple marks, 240 million years old—Moenkopi formation.*

Created, Designed, and Published in the U.S.A.
Printed by Doosan Dong-A Co., Ltd., Seoul, Korea
Paper produced exclusively by Hankuk Paper Mfg. Co., Ltd.